Asteroid Hitting Earth:

Understanding the Impact and What You Need to Know.

By

Eller J. Yazzie

This publication's content is fully protected by copyright law. Reproduction, distribution, or transmission in any form or by any means, including photocopying, recording, or any electronic or mechanical methods, is strictly prohibited without prior written permission from the publisher. Short quotes may be used in reviews or for certain noncommercial purposes permitted by copyright law. Any unauthorized use or reproduction violates the copyright holder's rights.

Copyright © Eller J. Yazzie, 2024.

Table of Contents

Introduction: Understanding Asteroids and Their Impact..........................7

Chapter 1: What Are Asteroids?...................16

Chapter 2: How Do Asteroids Travel in Space?........................... 27

- What Makes an Asteroid Move Through Space?...................................27
- How Do Asteroids Get Close to Earth?..29
- Can We Track Them Before They Arrive?...................................31
- Why is Tracking Asteroids Important?...35
- Conclusion... 36

Chapter 3: The Science Behind Asteroid Impacts... 38

- What Happens When an Asteroid Hits Earth?..38
- How Does the Earth's Atmosphere Affect Asteroids?......................................41
- What Causes Explosions When Asteroids Fall?..43
- The Aftermath of an Asteroid Impact. 46
- Conclusion... 48

Chapter 4: Famous Asteroid Impacts in History... 50

- The Dinosaur Extinction Event...........50
- Tunguska Event in Russia (1908)....... 53
- Other Notable Asteroid Impacts in Earth's Past... 56
- Why Studying Asteroid Impacts is Important...59
- Conclusion... 61

Chapter 5: What Could Happen If an Asteroid Hit Today?...62

- How Big Does an Asteroid Need to Be to Cause Damage?................................... 62
- Would an Asteroid Destroy a City or the Whole Planet?...65
- What Would Happen to Animals and Plants?... 68
- What Are We Doing to Protect Ourselves?... 71
- Conclusion.. 73

Chapter 6: How Do Scientists Track Asteroids?....................................... 74

- How Do Space Agencies Find Asteroids in Space?...74
- Tools and Technology Used to Spot Asteroids..77

- How Early Can We Spot a Dangerous Asteroid?... 80
- What Happens After an Asteroid is Found?.. 83
- Conclusion... 84

Chapter 7: Can We Prevent an Asteroid Impact?...86

- What Plans Do Scientists Have to Stop an Asteroid?...86
- How Could We Change an Asteroid's Path?.. 88
- Is There a Way to Break Up an Asteroid?... 91
- What If We Can't Stop the Asteroid?. 93
- Conclusion... 95

Chapter 8: How Do Space Agencies Prepare for Asteroid Impacts?...97

- What Do Space Agencies Do to Stay Safe from Asteroids?.............................. 97
- The Role of NASA and the European Space Agency.. 99
- What Happens if an Asteroid is Spotted Close to Earth?.. 102
- International Cooperation in Protecting Earth... 105
- Conclusion.. 106

Chapter 9: What Can We Do to Stay Safe?..... 108

Should We Worry About Asteroid Impacts?...108

What Precautions Can People Take?.... 110

How Do Scientists Inform the Public About Risks?.. 114

Conclusion... 118

Chapter 10: The Future of Asteroid Impact Studies.............120

What New Technologies Are Being Developed to Track Asteroids?............ 120

How Will Scientists Learn More About Asteroids in the Future?....................... 125

What Is the Long-Term Goal of Asteroid Impact Prevention?............................129

Conclusion..133

Conclusion: Why We Must Keep Learning About Asteroids............................. 134

The Importance of Continued Asteroid Research... 134

How Understanding Asteroids Can Keep Us Safe... 136

What You Can Do to Stay Informed.... 139

Conclusion: Why It Matters................. 142

Introduction: Understanding Asteroids and Their Impact

- When we look up at the night sky, we often see stars and the moon, but there's something else out there too: asteroids. These space rocks may look small from a distance, but they are very important. Some of them are big enough to cause big changes on Earth if they hit us. That's why understanding asteroids is so important. This chapter will explain what asteroids are, why they matter, and how we can learn more about them. You'll also learn why it's

important to study their impact and how it affects our planet.
- What is an asteroid?
- An asteroid is a large rock or chunk of metal that travels through space. They are smaller than planets but bigger than grains of dust. Most asteroids are found in the asteroid belt, which is a space between Mars and Jupiter. But not all asteroids stay in the asteroid belt. Some break away and travel toward Earth.
- There are many different kinds of asteroids. Some are very small, only a few meters wide, while others can be as big as a mountain. Even though they are made of rock or metal, they are not always the same. Some asteroids are made of metal, while

others are made of a mix of rock and ice. Most of these space rocks are very old, many billions of years old, and they are like pieces of early space history.
- Why do we care about asteroids hitting Earth?
- Even though asteroids are far away in space, they are important for Earth. Some of them can be dangerous if they hit our planet. When an asteroid comes too close to Earth, it could cause damage to the land, oceans, and even the air. Depending on the size of the asteroid, the impact could be small, like creating a crater, or very large, causing big changes to our world. If a large asteroid were to hit, it

could destroy cities, cause wildfires, or even change the Earth's climate.
- While big asteroid impacts are rare, smaller ones happen more often. Most of these smaller asteroids burn up when they enter the Earth's atmosphere, so they don't cause damage. However, scientists know that there's a chance of a bigger asteroid hitting in the future, and this is why we must study them. Understanding asteroids can help us prepare for potential impacts and prevent or reduce the damage they might cause.
- How can we learn about these space rocks?
- Scientists study asteroids using powerful tools. One of the most

important tools is telescopes. With telescopes, scientists can look far into space and spot asteroids that are on their way toward Earth. There are many space agencies around the world, like NASA and the European Space Agency, that use these telescopes to monitor the skies for any dangerous asteroids.

- In addition to telescopes, scientists also send special spacecraft to study asteroids up close. These spacecraft can land on the surface of an asteroid to learn about its size, shape, and what it is made of. By studying asteroids this way, scientists can get a better idea of where they are headed and whether or not they might pose a threat to Earth. They can also collect

samples from the asteroid, bringing pieces of it back to Earth for closer study.
- By using these tools, scientists can track asteroids that are on a path toward Earth. When they spot one, they can calculate how close it will come and if it is in danger of hitting us. Sometimes, they can predict these impacts years or even decades in advance.
- Why is it important to understand their impact?
- Understanding how an asteroid might impact Earth is very important. If we know that a large asteroid is coming, we need to prepare for it. Scientists study what happens when an asteroid hits to better understand how much

damage it could cause. This can help them come up with ways to protect Earth.

- When an asteroid hits Earth, it can cause a huge explosion. The energy from the explosion can cause fires, earthquakes, and even tsunamis, or giant waves in the ocean. The impact can also send dust and gas into the air, blocking out sunlight and making the planet colder. This can lead to a "nuclear winter" effect, where temperatures drop for a long time, making it hard for plants and animals to survive.
- Knowing this, scientists are working on ways to stop or slow down dangerous asteroids. They are looking for ways to push asteroids off course

or even destroy them before they can hit. This kind of work takes a lot of time, money, and effort, but it could save lives in the future. By understanding how asteroid impacts happen, we can better protect ourselves and our planet.
- Studying asteroid impacts also helps us learn more about the history of Earth. Many scientists believe that a big asteroid impact 66 million years ago caused the extinction of the dinosaurs. By understanding how this event happened, we can learn about the Earth's past and how life on our planet has changed over time.
- In conclusion, understanding asteroids and their impact is very important for our future. Asteroids

might seem like small, distant rocks, but they can have a big effect on Earth. By studying them, we can learn how to spot dangerous ones and prepare for their impact. Scientists are working hard to protect us from potential asteroid threats, and by learning more about these space rocks, we can keep our planet safe. The more we know, the better we can understand how to live alongside these space rocks without worrying about them causing harm.

Chapter 1: What Are Asteroids?

Asteroids are space rocks that float in space, traveling around the Sun. They can be big or small, made of different materials, and have different shapes. But, how do they differ from other space objects, like comets and meteoroids? In this chapter, we'll look at what asteroids are, where they come from, how big they can be, and the different types of asteroids that exist.

The Difference Between Asteroids, Comets, and Meteoroids

At first, it might seem like asteroids, comets, and meteoroids are the same thing, but they are very different from each other.

- Asteroids are made of rock and metal. They are larger than meteoroids and don't have tails. Most asteroids are found in the asteroid belt, which is a region of space between the planets Mars and Jupiter.

- Comets are made of ice, dust, and rock. They are known for their beautiful, glowing tails. When a comet gets close to the Sun, the heat causes the ice to melt, and the comet starts to release gas and dust, forming a long, bright tail that stretches across the sky.

- Meteoroids are smaller than asteroids and are made of rock or metal. When a meteoroid enters Earth's atmosphere and burns up, it creates a bright flash of light called a meteor, or "shooting star." If the meteoroid is big enough to hit the Earth's surface, it is called a meteorite.

In short, the main difference is that asteroids are larger, made mostly of rock and metal, and don't have tails like comets. Meteoroids are smaller pieces of rock or metal that travel through space.

Where Do Asteroids Come From?

Asteroids mostly come from the asteroid belt, a region in space between the planets Mars and Jupiter. This belt is full of thousands of asteroids of different sizes. Scientists believe that the asteroid belt is made up of leftover pieces from the early solar system. A long time ago, when the planets were forming, these pieces of rock and metal didn't come together to form a planet. Instead, they stayed in the asteroid belt, orbiting around the Sun.

Sometimes, an asteroid can be pushed out of the belt by the gravity of other planets, like Jupiter. When this happens, the asteroid can travel toward Earth or other planets. These asteroids can then become Near-Earth Objects (NEOs), which are asteroids that come close to Earth's path.

There are also some asteroids that come from farther away. Some come from the Kuiper Belt, which is beyond Neptune, and others come from a region called the Oort Cloud, which is much farther out in space.

Asteroids don't just stay in one place forever. They can change their paths and sometimes crash into each other, creating smaller pieces that float through space. These broken pieces can become new asteroids or meteoroids.

How Big Are Asteroids?

Asteroids can be many different sizes. Some are very small, like the size of a pebble, while others are much, much bigger, like the size of a mountain. In fact, the biggest asteroid we know about is Ceres, which is also classified as a dwarf planet. Ceres is around 590 miles (940 kilometers) across and is much larger than most other asteroids.

Most asteroids, however, are smaller than Ceres. They can be anywhere from a few feet wide to several miles wide. An asteroid that is about 330 feet (100 meters) across is big enough to cause some damage if it hits Earth, but it's still much smaller than Ceres. The size of an asteroid affects how much damage it could cause if it hit Earth. A small asteroid would burn up in Earth's atmosphere, while a bigger one could cause huge damage to the ground.

In general, asteroids are much smaller than planets, but they are still big enough to be dangerous if they hit Earth. Some asteroids have even caused huge impacts in the past, leaving craters on Earth's surface.

What Are the Different Types of Asteroids?

There are different types of asteroids, and they are categorized based on their composition—what they are made of. The three main types of asteroids are:

1. C-type (Carbonaceous) Asteroids

    These are the most common type of asteroids, making up about 75% of all known asteroids. C-type asteroids are made mostly of carbon, along with rock and minerals. They are dark and reflect very little sunlight, which is why they are hard to see from Earth. These asteroids are found mostly in the outer part of the asteroid belt.

2. S-type (Silicaceous) Asteroids

    S-type asteroids are made mostly of

silicate rock, which contains minerals like silicon and oxygen. They are much brighter than C-type asteroids and can be seen more easily. S-type asteroids are typically found in the inner part of the asteroid belt, closer to Mars.

3. M-type (Metallic) Asteroids

M-type asteroids are made mostly of metals, such as iron and nickel. These asteroids are shiny and reflect more sunlight than the other types. They are rarer than the C-type and S-type asteroids, but they are very important because scientists believe they might contain valuable metals. M-type asteroids are often found in the middle of the asteroid belt.

Besides these three main types, there are other less common types of asteroids. For example, some asteroids are made of ice, while others might have a mixture of different materials. Scientists are always learning more about asteroids and their different types, which is important for understanding how they move and how they might affect Earth.

Asteroids and Their Importance

Understanding asteroids is important for many reasons. First, learning about their size and composition helps scientists figure out how dangerous they could be if they ever hit Earth. By studying the types of asteroids and how they travel through space, scientists can better track their paths and predict if any of them might get too close to our planet.

Also, asteroids can teach us about the early solar system. Since they are made of leftover materials from when the solar system formed, they hold clues about the building blocks of the planets. By studying asteroids, scientists can learn more about how Earth and the other planets were made billions of years ago.

Finally, some asteroids might be valuable. They can contain metals and minerals that are rare on Earth. In the future, scientists might be able to mine asteroids for these materials, which could be useful for space missions or for use here on Earth.

Conclusion

In this chapter, we've learned that asteroids are space rocks that come in many different shapes, sizes, and materials. They are very important to study because they can help us understand the early solar system, protect Earth from potential dangers, and even provide valuable resources. Now that you know the basics of what asteroids are, their different types, and where they come from, you can better understand why scientists spend so much time studying them. In the next chapter, we will explore how asteroids travel in space and what happens when they get close to Earth.

Chapter 2: How Do Asteroids Travel in Space?

- Asteroids are always moving through space. They don't stay still. Just like Earth orbits the Sun, asteroids travel in their own paths around the Sun, sometimes getting very close to Earth and other planets. Understanding how asteroids move helps scientists know if they might come near Earth. Let's look at how asteroids move through space, how they can get close to Earth, and if we can track them before they arrive.
- What Makes an Asteroid Move Through Space?
- Asteroids move through space because of the force of gravity. Gravity is an

invisible force that pulls objects toward each other. The Sun's gravity pulls everything in the solar system, including asteroids, planets, and comets. This is what keeps the planets in their orbits and keeps asteroids moving through space in a set path.

- Asteroids travel around the Sun in what's called an orbit. An orbit is the path that an object takes as it moves around a bigger object. Just like the Earth orbits around the Sun, asteroids orbit around the Sun too. The Sun's gravity keeps them moving in their orbits, but asteroids don't always follow the same path. Their orbits can be different shapes, sometimes oval-shaped or more circular, depending on where they come from.

- Most asteroids are found in the asteroid belt, which is between the planets Mars and Jupiter. They are constantly moving along this belt, but their paths can change. Sometimes, asteroids might get knocked off their path by other objects in space, like when two asteroids crash into each other. These crashes can send asteroids into new orbits, causing them to move closer or farther from Earth.
- How Do Asteroids Get Close to Earth?
- Asteroids can get close to Earth if their orbits change. When an asteroid's path crosses Earth's orbit, it can come close to our planet. This happens because the gravity of other planets, like Jupiter, can push asteroids out of

their usual orbit. Jupiter has a very strong gravity, and sometimes it can pull on an asteroid, changing its direction and pushing it closer to Earth.

- When an asteroid comes close to Earth, it is called a Near-Earth Object (NEO). NEOs are asteroids or comets that are near enough to Earth to be noticed by scientists. Some NEOs pass close to Earth without hitting it, while others might get even closer. In some cases, an asteroid could even collide with Earth, but this is rare. Most asteroids that come close to Earth are small and harmless. Still, scientists keep track of these objects because it's important to know if any of them might be a danger.

- It's not only Jupiter's gravity that can cause an asteroid to get close to Earth. Other things can also change an asteroid's orbit. Sometimes, an asteroid's own gravity or the gravity of other objects in space can pull it closer to Earth. Even the heat from the Sun can push an asteroid to move in different directions. As the Sun heats up one side of an asteroid, it can cause the asteroid to spin and change its path, making it come closer to Earth.
- Can We Track Them Before They Arrive?
- Yes, scientists can track asteroids before they arrive, and they do this all the time. In fact, there are special programs and telescopes designed to spot asteroids early and follow their movements through space. Tracking

asteroids is very important because it allows scientists to know if any of them might get too close to Earth.

- The process of tracking asteroids starts with telescopes. Telescopes are powerful tools that help scientists see objects in space, even if they are very far away. There are special telescopes around the world and in space that are used to look for asteroids and comets. These telescopes can see asteroids far before they reach Earth, sometimes even years or decades in advance.
- When an asteroid is spotted, scientists will track its movement over time. They do this by watching how the asteroid moves across the sky and recording its position. By watching it for a long period, scientists can figure

out where the asteroid is going and if it might come too close to Earth. They also study the asteroid's speed, size, and shape to get a better idea of how it will behave.

- There are also space missions that have been sent out to learn more about asteroids. For example, NASA has sent missions like OSIRIS-REx to study asteroids up close. These missions can take photos, collect samples, and help scientists understand more about the asteroid's path and behavior. By learning about the asteroid's size and speed, scientists can make better predictions about where it will go.
- In addition to telescopes and space missions, scientists also use

computers to help track asteroids. By entering information about the asteroid's orbit and speed, computers can make models that show where the asteroid will be in the future. This allows scientists to predict how close an asteroid will get to Earth, and if there is any chance that it could collide with our planet.

- One important thing to note is that not all asteroids can be tracked. Some asteroids are small and hard to spot, especially if they are far from Earth or moving fast. However, thanks to new technology, we are finding more and more asteroids and tracking their movements with greater accuracy. Space agencies, like NASA and the European Space Agency (ESA), work

together to keep an eye on asteroids and warn people if one is coming too close.
- Why is Tracking Asteroids Important?
- Tracking asteroids is very important because it helps keep us safe. If a large asteroid were to hit Earth, it could cause a lot of damage. While this is very rare, it's better to be prepared and know if an asteroid is coming. By spotting an asteroid early, scientists have more time to figure out how to protect Earth.
- In the past, there have been examples of asteroids hitting Earth. Some of these impacts caused big craters, like the one in Mexico that helped cause the extinction of the dinosaurs. But today, thanks to new technology, we

have a better chance of spotting and tracking these objects.

- In some cases, scientists might be able to do something about an asteroid before it reaches Earth. If we know an asteroid is coming, we might be able to change its path using special technology. This is still being studied, but it's one of the ways scientists are thinking about protecting Earth from potential asteroid impacts.
- Conclusion
- In this chapter, we've learned how asteroids travel through space, how they can get close to Earth, and how scientists track them. Asteroids move because of gravity, and their paths can change when they are affected by the gravity of other planets or objects in

space. We can track asteroids using telescopes, computers, and even space missions. By doing so, scientists can predict if an asteroid will get too close to Earth and work on ways to protect our planet. In the next chapter, we will explore what happens when an asteroid hits Earth and the potential effects it could have on our planet.

Chapter 3: The Science Behind Asteroid Impacts

- Asteroids are fascinating space rocks, but when they hit Earth, the results can be quite serious. The idea of an asteroid hitting Earth can sound scary, but scientists study how these impacts happen so that we can better understand what they mean for our planet. In this chapter, we will learn what happens when an asteroid hits Earth, how Earth's atmosphere affects asteroids, and why explosions happen when they fall.
- What Happens When an Asteroid Hits Earth?

- When an asteroid comes into contact with Earth, it can cause a lot of damage, depending on how big it is. The impact creates an explosion that releases a huge amount of energy. If the asteroid is big enough, it can cause a crater to form on the Earth's surface. The size of the crater depends on the size and speed of the asteroid.
- For smaller asteroids, the damage is usually less severe. The rock might burn up in the Earth's atmosphere before it even reaches the ground. However, if the asteroid is large enough to reach the surface, it can cause a lot of destruction. The impact can create shock waves that spread out, shaking the ground and causing

buildings to collapse or even triggering earthquakes.
- Large asteroids, especially those over a kilometer wide, can cause even more serious damage. They can create massive explosions that send debris flying everywhere. This can create fires, trigger tsunamis if they land in oceans, and even change the climate. The explosion might throw dust and debris into the sky, blocking out sunlight and causing a global cooling effect that could last for years. These kinds of asteroid impacts are very rare, but they can have dramatic effects on life on Earth.
- The biggest asteroid impacts in Earth's history have been linked to the extinction of species. For example, the

asteroid that struck Earth millions of years ago is believed to have contributed to the extinction of the dinosaurs. When such large asteroids hit Earth, the effects can be felt all over the planet, and life can be greatly affected.

- How Does the Earth's Atmosphere Affect Asteroids?
- Earth's atmosphere plays a very important role when it comes to asteroid impacts. The atmosphere is the layer of gases that surrounds Earth. It acts like a shield, protecting the planet from many space objects, including asteroids. When an asteroid enters Earth's atmosphere, it starts to slow down because of the friction with the air. This friction causes the

asteroid to heat up, sometimes to very high temperatures.

- For small asteroids, the heat can be enough to cause them to burn up completely before they ever hit the ground. These small rocks, known as meteoroids, can create bright streaks of light in the sky, called meteors. When you see a shooting star, you are actually seeing a meteoroid burn up in the atmosphere. These meteors can be very small, and by the time they reach the Earth's surface, they are often completely gone.
- For larger asteroids, the atmosphere still helps by slowing them down and burning them up to some extent. However, if the asteroid is large enough, it will survive its trip through

the atmosphere and crash onto the Earth's surface. When this happens, the asteroid can cause a big explosion, especially if it is moving very fast.

- Most asteroids that hit Earth are slowed down and partially burned up by the atmosphere. Still, larger ones can make it through the atmosphere, and that is when we see the biggest impacts. Scientists closely monitor the paths of these large asteroids to understand how much danger they could pose to Earth.
- What Causes Explosions When Asteroids Fall?
- When an asteroid hits the Earth's surface, it doesn't just fall gently. The speed at which the asteroid travels through space is incredible. Asteroids

can travel at speeds of up to 25 kilometers per second (about 56,000 miles per hour). This is much faster than the speed of sound or any car on the road. Because of this high speed, the energy released when the asteroid hits the ground is massive.

- When the asteroid hits Earth, all that speed turns into energy, and this energy gets released in the form of an explosion. The explosion happens because of the kinetic energy the asteroid has. Kinetic energy is the energy an object has because of its motion. When an asteroid moves at such high speeds, the amount of energy it carries is very large. When the asteroid suddenly comes to a stop upon hitting the Earth, all that energy

has to go somewhere. The energy is released as heat, light, and shock waves. This is what causes the explosion.
- The explosion from an asteroid impact is very powerful. It can break apart rocks and soil, send pieces flying in every direction, and create fires. The explosion can also cause the Earth's surface to shake. This shaking, called seismic activity, can lead to earthquakes or even cause volcanic eruptions.
- In addition to the initial explosion, the impact can also cause a shock wave, which is a type of pressure wave that travels through the air. The shock wave can be felt miles away from the impact site and can cause damage to

buildings and structures. This is why it's important to track asteroids and understand how much damage they could cause.

- When large asteroids hit Earth, the explosion can be so big that it affects the entire planet. For example, a large asteroid impact can cause wildfires, especially if the explosion releases heat into the atmosphere. The heat can also cause large amounts of dust and smoke to rise up into the air. This dust can block sunlight, leading to cooler temperatures on Earth. This cooling effect can last for months or even years, and it is one reason why large asteroid impacts in the past have caused mass extinctions.
- The Aftermath of an Asteroid Impact

- After an asteroid explodes, the effects can be felt for a long time. If the asteroid is large enough, it can cause what is called global cooling. This happens because the explosion sends dust and particles into the atmosphere, which blocks out sunlight. Without sunlight, the Earth's temperature drops, and this can affect the climate for a long time.
- In addition to global cooling, the impact can create tsunamis. If an asteroid strikes an ocean or sea, the explosion can send huge waves of water crashing onto land, causing flooding and destruction. These waves can be very dangerous and can spread across the globe.

- Asteroid impacts can also affect the air quality. The explosion can send chemicals and gases into the atmosphere, which can change the composition of the air we breathe. Some of these gases can even affect plants and animals, making it harder for life to survive after a big impact.
- Conclusion
- In this chapter, we've learned about what happens when an asteroid hits Earth, how the atmosphere affects asteroids, and why explosions occur when asteroids fall. The speed of the asteroid, the energy it carries, and the friction it faces in the atmosphere all play a role in the explosion that occurs during an impact. These explosions can cause a lot of damage, from

craters to fires to earthquakes. Understanding these impacts is important because it helps scientists track asteroids and understand how they can affect Earth. In the next chapter, we will explore how scientists study asteroid impacts and work to protect Earth from potential dangers.

Chapter 4: Famous Asteroid Impacts in History

- Throughout Earth's long history, there have been several famous asteroid impacts that have shaped the planet and its life. Some of these impacts were so powerful that they caused major changes, even leading to the extinction of species. In this chapter, we will look at some of the most famous asteroid impacts: the event that wiped out the dinosaurs, the Tunguska event in Russia in 1908, and other notable asteroid impacts from Earth's past.
- The Dinosaur Extinction Event

- One of the most well-known asteroid impacts in history is the one that happened around 66 million years ago. This event is believed to have caused the extinction of the dinosaurs. The asteroid that struck Earth at that time was enormous, about 10 kilometers (6 miles) wide. It hit what is now the Yucatán Peninsula in Mexico, creating a massive crater called Chicxulub.
- When the asteroid hit the Earth, it released an incredible amount of energy. The explosion was so powerful that it caused huge wildfires, and the impact sent massive amounts of dust and debris into the atmosphere. This dust blocked the sunlight, leading to a dramatic drop in temperatures around

the world. Without sunlight, plants couldn't grow, and this affected the food chain. Animals that depended on plants and smaller animals for food were also affected, leading to a collapse of ecosystems.

- This sudden change in the climate, known as a nuclear winter, lasted for months or even years. The lack of sunlight and the cooling of the Earth caused many species of plants and animals to die. The dinosaurs, who were at the top of the food chain, couldn't survive these harsh conditions. As a result, they went extinct. This mass extinction event is one of the most significant in Earth's history, and it is linked to the asteroid impact at Chicxulub.

- Scientists have studied the evidence left behind by the asteroid impact. There are layers of rocks around the world that show high levels of iridium, a rare metal often found in asteroids. These layers help scientists confirm that an asteroid impact was the cause of the mass extinction event. While the exact cause of the dinosaurs' extinction is still debated, the asteroid impact remains the most widely accepted theory.
- Tunguska Event in Russia (1908)
- Another famous asteroid event in history happened in 1908, in a remote area of Siberia, Russia. This event, known as the Tunguska event, was caused by a space rock, possibly a comet or an asteroid, that exploded in

the atmosphere. The object was estimated to be about 50 to 60 meters (160 to 200 feet) in size. It exploded with the force of about 10 to 15 megatons of TNT, which is around 1,000 times stronger than the atomic bomb dropped on Hiroshima during World War II.

- Even though the object did not hit the Earth's surface, the explosion created a shockwave that flattened around 2,000 square kilometers (770 square miles) of forest. Thousands of trees were knocked down, and the explosion was felt hundreds of kilometers away. People who were in the area saw a bright flash in the sky, followed by a loud bang. In the days that followed, the sky was filled with smoke and

dust, which blocked the sun and caused strange weather patterns.
- The Tunguska event is important because it showed just how powerful an asteroid or comet can be, even if it doesn't hit the ground. The explosion caused a lot of damage, but luckily, the area where it occurred was sparsely populated, so there were no known human casualties. However, this event raised awareness about the potential danger that space objects pose to Earth.
- Scientists believe that the object that caused the Tunguska event was made of ice and rock, similar to a comet. It entered the atmosphere at a very high speed and exploded before it reached the ground. The explosion released

energy in the form of heat and pressure, causing trees to be knocked down and creating a fireball that could be seen from far away.
- Today, the Tunguska event is still studied by scientists, who are trying to understand how objects like this can impact the Earth. Although the explosion occurred more than a century ago, it serves as a reminder of the power of asteroids and comets and the importance of monitoring objects that come near our planet.
- Other Notable Asteroid Impacts in Earth's Past
- While the asteroid that caused the dinosaur extinction event and the Tunguska event are among the most famous, they are not the only asteroid

impacts that have shaped Earth's history. There have been many other asteroid impacts throughout the planet's past, some of which have caused significant changes to Earth and its ecosystems.

- One such impact occurred around 3.5 billion years ago. This impact, known as the Late Heavy Bombardment, was a period during which Earth was hit by a large number of asteroids and comets. Scientists believe that this bombardment played a major role in shaping the early Earth and possibly even in the formation of the Moon. During this time, the Earth was constantly being hit by space rocks, and the impacts created craters and caused volcanic activity. The Late

Heavy Bombardment is believed to have played a role in the formation of the Earth's crust and oceans.

- Another notable impact occurred about 35 million years ago, when an asteroid struck what is now known as the Chesapeake Bay area in the United States. This impact created a crater that is about 85 kilometers (53 miles) wide. The impact released enough energy to cause a shockwave, and it is thought to have created massive tsunamis in the surrounding area. Although the event did not lead to an extinction, it still caused significant changes in the local environment.
- In addition to these impacts, there have been many smaller impacts over the years that have left craters on

Earth. Some of these craters are easy to see, while others are buried beneath the Earth's surface. Scientists continue to study these craters to learn more about how asteroids have shaped the Earth's history and how they might impact the planet in the future.

- Why Studying Asteroid Impacts is Important
- Studying asteroid impacts is important for several reasons. First, it helps scientists understand how asteroids affect Earth and its ecosystems. By studying past impacts, scientists can learn how these events have shaped the planet over time. This information can also help scientists predict the effects of future asteroid impacts.

- Second, understanding asteroid impacts helps scientists develop ways to protect Earth from potential dangers. While large asteroid impacts are rare, they can have catastrophic consequences. By tracking the paths of asteroids and comets, scientists can better predict when an impact might occur and how to prevent or mitigate the damage.
- Finally, studying asteroid impacts can teach us about the history of our solar system. Asteroids are like time capsules from the early days of the solar system. By studying the materials that make up asteroids, scientists can learn more about the formation of the planets and the history of our solar system.

- Conclusion
- In this chapter, we've learned about some of the most famous asteroid impacts in history, including the event that wiped out the dinosaurs, the Tunguska event in Russia, and other notable impacts. These events show just how powerful and destructive asteroids can be, and why it is important to study them. By understanding the effects of past impacts, scientists can better prepare for future ones and continue to learn about the history of our planet and solar system. In the next chapter, we will explore how scientists track asteroids and what steps are being taken to protect Earth from potential asteroid impacts.

Chapter 5: What Could Happen If an Asteroid Hit Today?

- Asteroids are space rocks that can travel through space at incredible speeds. If one of these asteroids were to hit Earth today, it could cause different types of damage depending on how big the asteroid is. In this chapter, we will explore how large an asteroid needs to be to cause harm, whether it could destroy a city or even the entire planet, and what effects it might have on animals and plants.
- How Big Does an Asteroid Need to Be to Cause Damage?
- Asteroids come in many different sizes, from tiny pebbles to giant rocks

that are several kilometers wide. The size of the asteroid is very important when thinking about how much damage it could do if it were to hit Earth.

- Small asteroids, which are usually less than 25 meters (82 feet) wide, would burn up in the atmosphere before they could reach the ground. These small asteroids are not big enough to cause serious damage. Sometimes, we see these small rocks as "shooting stars" or meteors in the sky. They burn up in the air and disappear without causing harm to the Earth.
- However, larger asteroids can cause significant damage. An asteroid about 100 meters (328 feet) wide could cause an explosion strong enough to

destroy a small town or a city. It would hit the Earth with a lot of force, causing a huge explosion, fires, and strong shockwaves. This would be similar to a powerful bomb going off. If the asteroid were to hit an ocean, it could create huge waves, called tsunamis, which could flood coastal areas.

- Asteroids that are much larger, more than 1 kilometer (0.6 miles) wide, could cause even more serious damage. These huge asteroids can cause massive destruction. They could destroy entire cities and cause massive earthquakes. In the worst-case scenario, they could even change the climate of the Earth, making it much

harder for plants and animals to survive.
- Scientists have studied past impacts to understand how big an asteroid needs to be to cause major damage. The asteroid that wiped out the dinosaurs, for example, was about 10 kilometers (6 miles) wide. This impact caused global changes in the atmosphere and led to the extinction of many species. Even though such huge asteroids are rare, scientists still monitor the skies to track any large objects that could pose a threat to Earth.
- Would an Asteroid Destroy a City or the Whole Planet?
- Whether an asteroid could destroy just a city or the whole planet depends on its size. Small asteroids will only

damage a small area. However, larger asteroids have the potential to cause much larger destruction.

- If an asteroid were about 1 kilometer wide, it could completely destroy a large city. The explosion from the impact would level buildings, cause fires, and leave a large crater in the ground. The heat and pressure from the impact would also cause huge shockwaves, which could cause even more destruction far from the point of impact.
- For example, an asteroid of this size could hit a place like New York City or London and destroy everything in its path. While this would be a terrible event, it would only affect the area around the impact. The rest of the

world would still be safe, though the atmosphere could change. This is because the dust and debris from the impact could block out the sunlight, leading to cooler temperatures for months or even years. This effect is known as a "nuclear winter." The drop in temperature would make it harder for plants to grow, which could harm the food chain.

- If an asteroid larger than 10 kilometers wide were to strike Earth, it could cause damage on a global scale. This type of impact could create enough dust and debris to block the sun's light for a long time, which would dramatically lower temperatures. Plants wouldn't be able to grow without sunlight, and many

animals and plants could die because they wouldn't have enough food. The cooling of the Earth could last for a long time, causing problems for life on the planet.
- In the worst-case scenario, if the asteroid were large enough, it could destroy life on Earth entirely, similar to what happened when the dinosaurs went extinct. This kind of event, however, is extremely rare. Large asteroids are very hard to detect, but scientists are working hard to track any potential threats.
- What Would Happen to Animals and Plants?
- If an asteroid were to hit Earth today, the impact would have a huge effect on animals and plants. The size of the

asteroid and where it hits would determine how bad the impact would be.

- For smaller impacts, animals and plants in the immediate area would suffer the most. For example, if an asteroid were to hit a city, all the animals and plants nearby would be destroyed by the explosion, the fire, and the shockwaves. Even animals and plants far from the impact could be affected by the changes in the atmosphere. After the explosion, a large amount of dust and smoke would be released into the sky, blocking sunlight and causing temperatures to drop. This sudden change in temperature would make it difficult

for many plants and animals to survive.
- If the asteroid were large enough, like the one that wiped out the dinosaurs, the effects would be felt all over the world. The dust in the atmosphere would block sunlight, and this would stop plants from being able to photosynthesize. Without plants to produce food, herbivores would not have anything to eat. This would then affect animals that eat herbivores, creating a ripple effect through the food chain.
- The cooling of the Earth would also cause major problems. In addition to not having enough food, animals and plants would have to survive in much colder temperatures. Many species

would not be able to adapt quickly enough, and they would die off. This would lead to large-scale extinction events.

- It's not just large impacts that can affect life on Earth. Even smaller asteroid impacts can have long-term effects on ecosystems. If an asteroid were to hit the ocean, for example, it could create large tsunamis that would flood coastal areas. This would harm the animals and plants that live in those areas. The changes in the environment could also disrupt migration patterns for animals that rely on specific weather conditions.
- What Are We Doing to Protect Ourselves?

- While an asteroid impact is a scary thought, scientists are working hard to monitor the skies for any potential threats. There are space programs and telescopes that are constantly looking for asteroids that might come close to Earth. By detecting asteroids early, scientists can figure out if they are dangerous. If a threat is found, there are ideas being studied about how we could stop or deflect an asteroid before it hits Earth. One possible method is sending a spacecraft to nudge the asteroid off its path, so it doesn't collide with Earth.
- In the meantime, we can learn from past impacts to understand the dangers of space rocks and keep working to protect our planet.

- Conclusion
- In this chapter, we explored the possible effects of an asteroid impact on Earth. We learned that the size of the asteroid plays a major role in determining how much damage it can cause. Small asteroids are less dangerous, while larger ones could destroy cities, and even larger ones could change the entire planet's climate. We also looked at how animals and plants could be affected by an impact. While large asteroid impacts are rare, scientists continue to monitor the skies to keep us safe.

Chapter 6: How Do Scientists Track Asteroids?

- Asteroids are rocky objects that travel through space. Some are small, while others are much larger. Scientists want to know where these asteroids are and if any of them could pose a danger to Earth. In this chapter, we'll explore how scientists track asteroids, the tools they use, and how early they can spot a dangerous asteroid before it reaches Earth.
- How Do Space Agencies Find Asteroids in Space?
- Space agencies like NASA and other organizations around the world work together to find asteroids that are in

space. These agencies use special telescopes and other technology to search the skies. The reason they do this is to find asteroids that could be dangerous and to track their movements.

- The process of finding asteroids begins with looking at the sky. The sky is so big and full of stars, planets, and other objects that it can be difficult to find these small space rocks. However, scientists know where to look and use powerful telescopes to help them spot even small asteroids. The more they look, the more they find.

- One of the ways scientists find asteroids is by using surveys, which are like big, organized "search parties" in the sky. These surveys scan the sky every night to look for moving objects. Since asteroids move through space, they appear as small dots that change position over time, making them different from stars, which stay in the same place. These surveys take lots of pictures, and then scientists look at the pictures carefully to see if they can spot any moving asteroids.

- Many space agencies around the world are working together to track asteroids. Some of the most famous asteroid-tracking projects include NASA's Near-Earth Object Observations (NEOO) Program, the European Space Agency's (ESA) Near-Earth Object (NEO) program, and the Pan-STARRS project in Hawaii. These projects focus on

finding asteroids that are close to Earth and could potentially cause harm.
- Tools and Technology Used to Spot Asteroids
- Scientists use several types of tools and technology to spot asteroids. These tools help scientists see far into space and detect objects that are very far away.
- One of the main tools used to find asteroids is a telescope. A telescope is a big machine that lets scientists see distant objects in space. There are different types of telescopes, but the most common ones used for asteroid hunting are optical telescopes, infrared telescopes, and radar telescopes.

- Optical telescopes work like regular binoculars or telescopes you might use to look at stars. These telescopes gather light from the sky, and scientists use the images to find moving objects, such as asteroids. They can spot asteroids by noticing tiny dots in the pictures that shift their positions over time.
- Infrared telescopes are a little different. They detect heat instead of visible light. When an asteroid is in space, it gives off heat, and infrared telescopes can see that heat. This helps scientists find asteroids even if they aren't visible with regular telescopes. These telescopes are especially useful because they can find

asteroids that are very dark or far away.
- Radar telescopes are used to bounce radio waves off objects in space. When the radio waves hit an asteroid, they bounce back to Earth. Scientists can measure how long it takes for the radio waves to return, which helps them figure out how far away the asteroid is and how fast it is moving. Radar telescopes also give scientists detailed pictures of asteroids, which can help them understand the asteroid's shape and size.
- In addition to telescopes, scientists use satellites and space probes to study asteroids. Satellites are machines that orbit Earth and can look out into space. Space probes are

spacecraft that travel to asteroids to study them up close. Some space probes have even landed on asteroids and taken pictures of the surface.
- Another important tool is computer software. Scientists use powerful computers to analyze all the data they collect from telescopes and other tools. This software helps them track the movement of asteroids, predict their paths, and see if any could come close to Earth.
- How Early Can We Spot a Dangerous Asteroid?
- Finding dangerous asteroids early is one of the most important jobs for scientists. The earlier they spot an asteroid, the more time they have to figure out how to deal with it. But how

early can we spot a dangerous asteroid?
- In many cases, scientists can spot asteroids several years before they might reach Earth. For smaller asteroids, scientists can sometimes spot them a few weeks or even days before impact. However, bigger asteroids are easier to find from a greater distance. If an asteroid is large enough to cause serious damage, scientists usually have more time to track it.
- For example, NASA's Near-Earth Object Observations (NEOO) Program has helped spot many asteroids years before they could come near Earth. Some asteroids are discovered many years in advance, and scientists can

carefully watch their movement. They can also make calculations to predict exactly where the asteroid will be in the future. This is very helpful because it gives people time to prepare if the asteroid is dangerous.

- In some cases, scientists can find an asteroid years or even decades before it might reach Earth. With this much time, they can carefully track the asteroid's path and study how it might change over time. They can also test ways to change the asteroid's course, using methods like sending a spacecraft to nudge it off course, so it doesn't hit Earth.

- However, there is still some risk. Even though scientists are always looking for new asteroids, some smaller asteroids might still slip by unnoticed. If an asteroid is smaller than 140 meters (459 feet), it is harder to spot because it doesn't reflect much light and might not be noticed until it gets very close to Earth. But scientists are working on improving technology to find even the smallest asteroids, so they can be ready if any of them are dangerous.

- What Happens After an Asteroid is Found?
- When an asteroid is found, scientists continue to track it carefully. They observe its movement, size, and speed. Using this information, they can predict whether the asteroid will collide with Earth or pass by safely. If an asteroid is going to pass by,

scientists can calculate how close it will come to Earth. They can even calculate if it will get too close and cause a risk.
- If an asteroid is found to be on a collision course with Earth, scientists have a few options. One option is to send a spacecraft to change its direction. This could be done by crashing a spacecraft into the asteroid, using its force to push the asteroid off its path. Another option is to use a special laser to vaporize part of the asteroid, changing its speed and direction.
- Conclusion
- In this chapter, we've learned about how scientists track asteroids and the technology they use to spot them.

Scientists use telescopes, satellites, radar, and space probes to find asteroids and study their movement. Thanks to these tools, we can spot dangerous asteroids years in advance. Tracking asteroids is an important job because it gives scientists time to prepare and protect Earth if necessary. With the help of new technology, scientists continue to make progress in finding and understanding asteroids, keeping our planet safe from space rocks.

Chapter 7: Can We Prevent an Asteroid Impact?

- Asteroids are huge rocks that move through space. Sometimes, these rocks can come close to Earth. If one of them hits our planet, it could cause big problems. Scientists work hard to understand asteroids and figure out ways to stop them if they are on a path to hit Earth. In this chapter, we will explore the plans scientists have to prevent an asteroid impact, how they could change an asteroid's path, and if it's possible to break up an asteroid before it hits Earth.
- What Plans Do Scientists Have to Stop an Asteroid?

- Preventing an asteroid impact is a big challenge, but scientists are always working on new ideas. The good news is that there are some plans already in place that could help us stop an asteroid from hitting Earth. These plans rely on using technology, understanding the asteroid's size, and predicting its path in space. By knowing where the asteroid is going and how big it is, scientists can figure out the best way to stop it.
- One of the main plans is to track asteroids in space. As we talked about in the previous chapter, space agencies like NASA are always looking for asteroids. They use powerful telescopes and other tools to spot asteroids that could be dangerous. If

they find one, they will track its path over time to see if it's heading toward Earth.
- If scientists discover an asteroid that could hit Earth, they will have some time to come up with a plan. They could use rockets or spacecraft to try to change the asteroid's path, so it doesn't collide with Earth. There are many ideas for how to do this, and scientists are working on which ones are the best and safest to use.
- How Could We Change an Asteroid's Path?
- Changing the path of an asteroid is one of the most important ways we could prevent a disaster. To do this, scientists need to move the asteroid just a little bit so it avoids hitting

Earth. It's not about making huge changes, but rather small adjustments that can have a big impact over time.

- One idea is to send a spacecraft to push the asteroid. This spacecraft could hit the asteroid, and the force from the impact would change the asteroid's speed and direction. Even a small change in speed could make the asteroid miss Earth by a wide margin. This method is called the "kinetic impactor" technique.
- Another idea is to use a spacecraft that doesn't crash into the asteroid but instead uses its engines to push the asteroid gently. This method is called the "gravity tractor" technique. The spacecraft would fly near the asteroid and use the force of its engines to pull

the asteroid in a different direction. This method would take more time because the spacecraft would need to stay near the asteroid for a long time, but it could be very effective.

- Scientists also think about using lasers to change the asteroid's path. If a laser is aimed at the asteroid, the heat from the laser could cause the asteroid to push away small amounts of material. This tiny force could slowly change the asteroid's direction. This method would need a very strong and powerful laser, but it could work well if there's enough time to use it before the asteroid gets too close.
- Another method involves using a spacecraft to plant explosives on the asteroid's surface. These explosives

could be designed to break the asteroid apart or push it in a different direction. This idea is still being researched and would require a lot of planning to make sure the explosion wouldn't cause more problems. It's important to be careful, as breaking the asteroid apart could send smaller pieces toward Earth.

- Is There a Way to Break Up an Asteroid?
- Breaking up an asteroid is another idea that scientists have thought about. If an asteroid is really big and would cause a lot of damage if it hit Earth, breaking it into smaller pieces could reduce the damage. Smaller pieces are less likely to cause as much harm, because they would burn up more easily in Earth's atmosphere.

- One way to break up an asteroid is by using explosions. A spacecraft could land on the asteroid, and scientists could place explosives on its surface. When the explosives are detonated, they would break the asteroid into smaller pieces. This could work well if the asteroid is made of softer materials. However, if the asteroid is very hard, the explosion might not be enough to break it up.
- Another idea for breaking up an asteroid is to send a spacecraft to blast it with a powerful laser. The laser could heat up parts of the asteroid until they crack or break off. This method would be safer than using explosives because it wouldn't involve physical contact with the asteroid, and

there would be no risk of creating a larger explosion. The downside is that this method would take a long time to work, and it might not be effective on every type of asteroid.

- Scientists have also looked into the idea of "nuclear" options. This means using the power of an atomic bomb to break up the asteroid. A nuclear bomb would release a huge amount of energy, which could break the asteroid apart. However, this option is risky. The pieces of the asteroid could still hit Earth, and it might create even more dangerous debris. Because of this, scientists would only consider using this method as a last resort.
- What If We Can't Stop the Asteroid?

- While there are many ideas for stopping an asteroid, it's important to remember that we might not always have enough time. If an asteroid is discovered very late, scientists might not be able to stop it in time. In these cases, it's important to have plans in place to help protect people and minimize damage.
- If we can't prevent the asteroid from hitting Earth, scientists are still working on ways to prepare for the impact. One plan is to create early warning systems. If we spot an asteroid early enough, people can be warned in advance, and governments can take steps to protect people. This might involve evacuating people from areas that are most likely to be hit or

building shelters to protect people from the blast.

- Scientists are also studying the effects of asteroid impacts. They want to understand what happens when an asteroid hits Earth, so they can better prepare. This includes studying how the impact could affect the environment, like causing fires or changes in the climate. By understanding these effects, scientists can help make plans to protect life on Earth.
- Conclusion
- In this chapter, we've learned about the different ways scientists are working to prevent an asteroid impact. There are several plans in place, like using spacecraft to push asteroids off

course or breaking them into smaller pieces. While it's not always easy, scientists are always working to improve the technology and ideas to stop dangerous asteroids. Even if we can't stop an asteroid, early warning systems and emergency plans can help protect people. The more we learn about asteroids and how to deal with them, the safer our planet will be.

Chapter 8: How Do Space Agencies Prepare for Asteroid Impacts?

- Space agencies around the world work hard to protect Earth from dangerous asteroids. These big space rocks can sometimes get too close to our planet, and if they hit, they could cause serious problems. In this chapter, we will learn how space agencies stay safe from asteroids, what NASA and the European Space Agency (ESA) do to help, and what happens if an asteroid is spotted close to Earth.
- What Do Space Agencies Do to Stay Safe from Asteroids?

- Space agencies are always looking for ways to protect Earth from the threat of asteroid impacts. Their main job is to watch the skies for asteroids and track their paths to make sure they won't hit Earth. They also work on plans to stop an asteroid if one is found to be heading our way.
- One of the most important things that space agencies do is to find and track asteroids. They use big telescopes and special tools to look for asteroids that could be dangerous. The earlier they spot an asteroid, the more time they have to figure out how to prevent it from hitting Earth. By studying the path of an asteroid, space agencies can predict where it will go and if it will get close to Earth.

- In addition to spotting asteroids, space agencies also try to learn more about them. They send spacecraft to study asteroids up close. This helps scientists understand what asteroids are made of and how they move. By learning more about asteroids, space agencies can come up with better ways to prevent them from hitting Earth.
- Space agencies also work together with other countries. They share information about asteroids and help each other keep track of any dangerous space rocks. This teamwork is important because it helps everyone be ready if an asteroid is found to be on a path toward Earth.
- The Role of NASA and the European Space Agency

- NASA is the space agency of the United States, and it plays a very important role in protecting Earth from asteroid impacts. NASA has a special program called the Near-Earth Object Program, which focuses on finding asteroids that are close to Earth. They use powerful telescopes to look for these objects, and they also work with other space agencies to share information.
- NASA has several projects aimed at learning more about asteroids. For example, NASA's OSIRIS-REx mission sent a spacecraft to an asteroid named Bennu. The spacecraft studied the asteroid and even collected samples to bring back to Earth. This mission will help scientists learn more about

asteroids and how to protect our planet from them.

- NASA also works on ways to prevent asteroid impacts. They have developed ideas like sending spacecraft to push an asteroid off course or using lasers to change its path. NASA also tests different methods to see which ones are the most effective in case they need to stop an asteroid in the future.

- The European Space Agency (ESA) also plays a key role in protecting Earth from asteroids. ESA works with NASA and other space agencies to track asteroids and share information. One of ESA's important projects is the Hera mission, which is part of an international effort to test how to stop an asteroid. Hera will visit two asteroids, one of which is part of a double asteroid system. The goal is to understand how an asteroid's path can be changed, which could help protect Earth if one is headed our way.

- ESA also runs a program called the Space Debris Office. This office tracks not only asteroids but also objects like old satellites and other space debris that could cause problems for spacecraft or Earth. ESA works on finding ways to keep space clean and safe for everyone.
- What Happens if an Asteroid is Spotted Close to Earth?
- If an asteroid is spotted close to Earth, space agencies immediately start working on a plan to deal with it. The first thing they do is confirm the

asteroid's size and path. They use telescopes to measure how big it is, how fast it is moving, and where it's going. This helps them figure out if the asteroid will hit Earth or if it will miss us.

- Once the asteroid's path is known, space agencies track it closely. They monitor its movement over time to make sure it's not getting closer. If the asteroid is going to hit Earth, the next step is to figure out the best way to stop it.
- In some cases, there might be enough time to change the asteroid's path. Space agencies could send a spacecraft to push the asteroid in a different direction, as we discussed in earlier chapters. If there isn't enough time to

push it, there are other options, such as breaking it into smaller pieces. These pieces would still be dangerous, but they would cause less damage if they hit Earth.

- Space agencies also make sure that people are prepared if an asteroid is about to hit. If there's no time to stop the asteroid, governments would need to take action. They could evacuate people from the area where the asteroid is going to land or build shelters to protect them. It's also possible that people would need to prepare for other dangers, like fires or changes in the climate, depending on the size and impact of the asteroid.
- Sometimes, when an asteroid is spotted, it may be difficult to stop it in

time. This is why early warning systems are so important. The more time space agencies have to prepare, the better they can protect people and reduce the damage. For this reason, space agencies are always working to improve their tracking systems and detection methods.
- International Cooperation in Protecting Earth
- One of the most important things about preparing for asteroid impacts is cooperation between countries. The threat of an asteroid is a global problem, and it requires teamwork from space agencies all over the world. NASA and ESA work together with other space agencies to share

information about asteroids and to plan for any potential threats.
- There are also international groups, like the United Nations, that work on asteroid impact prevention. These groups help to organize the efforts of different countries, so everyone is on the same page. By sharing knowledge and resources, countries can make sure they're ready to protect Earth from asteroid impacts.
- Conclusion
- In this chapter, we have learned how space agencies like NASA and the European Space Agency work together to protect Earth from asteroid impacts. They track asteroids using powerful tools, share information, and come up with plans to prevent

impacts. If an asteroid is spotted close to Earth, space agencies immediately start working on a plan to stop it. They could use spacecraft to change the asteroid's path or even break it into smaller pieces. International cooperation is crucial in protecting Earth from asteroids, and space agencies are always working to improve their methods to keep our planet safe.

Chapter 9: What Can We Do to Stay Safe?

Asteroids are big rocks that float in space. Some of them come close to Earth. While this might sound scary, scientists are working hard to make sure we stay safe. In this chapter, we will explore whether we should be worried about asteroid impacts, what precautions people can take, and how scientists tell everyone about the risks.

Should We Worry About Asteroid Impacts?

The first question many people have when they hear about asteroids is, "Should I be worried?" The truth is that while asteroid impacts are rare, they can cause big problems if they happen. However, scientists are constantly watching for asteroids that might come too close to Earth, so we don't need to worry every day. The chances of a big asteroid hitting Earth in our lifetime are very low.

Most asteroids that come close to Earth are small. These small asteroids burn up in the sky before they can reach the ground. Even though a few bigger ones have hit Earth in the past, space agencies around the world are using powerful tools to keep an eye on them. They track their paths to make sure that if a dangerous one is headed our way, we can act in time.

The idea of an asteroid hitting Earth is something that might make people feel scared. But it's important to know that scientists are very good at watching the skies and tracking asteroids. They work together across countries to protect Earth. They have a lot of information and technology to warn people in case something dangerous is found.

What Precautions Can People Take?

Even though the chances of an asteroid hitting Earth are very low, there are still some things people can do to stay safe. These are mainly things that scientists, governments, and space agencies are working on, but it's also important for everyone to know what could be done in an emergency.

1. Stay Informed: One of the best things people can do is stay informed about what is happening with asteroids.

Space agencies like NASA and the European Space Agency (ESA) track asteroids all the time. If an asteroid is found to be heading toward Earth, they will share that information with the public. By knowing what is going on, people can prepare if necessary.

2. Have an Emergency Plan: Just like with other natural disasters, it is good to have a plan in place in case something goes wrong. If a big asteroid is headed toward Earth, scientists would give plenty of notice. Governments would work to help people by telling them where to go and what to do. People might need to stay inside, especially if an asteroid is going to hit in a specific area. This

could help reduce the damage caused by the impact.

3. Building Safe Shelters: In the event that an asteroid impact is coming, building shelters could help protect people. Shelters can be built underground or in places that can withstand big impacts. For example, some buildings are made with strong walls and roofs to protect people during storms or other emergencies. In the case of a big asteroid impact, similar safe areas would be needed to protect people from flying debris or fires.

4. Evacuate the Area: If a large asteroid is about to hit a specific area,

evacuation might be necessary. Governments would give warnings and tell people where to go to be safe. They might direct people to other cities or regions where the impact will not be as dangerous. Evacuating the area is an important step to take if the asteroid's path is confirmed and the risk of harm is high.

5. Preparing for Changes: Some asteroids can cause changes to the environment. For example, if an asteroid is very large, it could send dust and particles into the air. This can block sunlight and affect the climate. If an asteroid were to hit, it could also cause fires and earthquakes. People can prepare for

these changes by knowing how to stay safe in case of fires, floods, or other dangerous events.

How Do Scientists Inform the Public About Risks?

When scientists discover an asteroid that might hit Earth, they do everything they can to inform the public. They use different ways to make sure everyone knows what is happening and how to stay safe.

1. Using the Media: One of the main ways that scientists share information with the public is through the media. TV channels, newspapers, and websites are all used to give updates about asteroids. If a dangerous

asteroid is spotted, scientists will work with news agencies to make sure people know. They will explain the size of the asteroid, when it might hit, and what to do.

2. Social Media: In today's world, social media is another important tool for sharing information. Space agencies like NASA use Twitter, Facebook, and other social media platforms to give real-time updates about asteroid risks. If something happens with an asteroid, these platforms are used to let people know right away.

3. Emergency Alerts: Governments also have systems in place to send emergency alerts to people. These

alerts can be sent through text messages, phone calls, and public loudspeakers. If an asteroid impact is going to happen soon, these alerts can warn people and help them get to safety quickly. For example, people might get a message telling them to go to a shelter or evacuate the area.

4. Websites and Apps: Space agencies also have websites and apps that people can check for up-to-date information. NASA's website, for example, has a page dedicated to asteroid tracking. People can use these resources to learn about asteroids and see if any are headed toward Earth. If you want to know if an asteroid is coming, these tools can give you all the

facts you need.

5. Working with Governments: When there's a risk of an asteroid impact, governments are involved in spreading the word. Space agencies share their findings with national governments, and the government will make announcements about what to do. They might send out emergency messages, set up evacuation routes, and make sure people are safe. Government leaders will also speak to the public about how to prepare and what to expect.

6. Educating the Public: Scientists are also working hard to educate people about asteroids. They create

educational programs, videos, and resources to teach everyone about the importance of asteroid tracking and what to do in an emergency. By teaching children and adults about asteroids, people will know more about the risks and how to stay safe.

Conclusion

While asteroid impacts are rare, it is still important for people to know how to stay safe. Scientists are constantly working to track asteroids and inform the public about potential risks. People should stay informed through the media, social media, and emergency alerts. Having an emergency plan, knowing how to prepare, and being ready for possible changes in the environment are all key ways to stay safe. If an asteroid is found to be on a path toward Earth, space agencies will inform the public and work with governments to make sure everyone is protected.

It's important to remember that, even though asteroid impacts are a real threat, the chances of one happening soon are very low. Thanks to the hard work of scientists and space agencies, we can rest easy knowing that they are keeping an eye on the skies and making sure Earth stays safe.

Chapter 10: The Future of Asteroid Impact Studies

As technology and science improve, our understanding of asteroids and how to protect Earth from possible impacts is getting better. In this chapter, we will look at the new technologies being developed to track asteroids, how scientists will learn more about them in the future, and the long-term goal of preventing asteroid impacts.

What New Technologies Are Being Developed to Track Asteroids?

To keep Earth safe from asteroids, scientists are constantly improving the tools and technology they use to track them. These new technologies help them spot asteroids earlier and more accurately. Some of the most exciting developments in asteroid tracking include:

1. Better Telescopes: Scientists use powerful telescopes to look at space and find asteroids. Some new telescopes are being built to be even stronger and clearer. These telescopes can help scientists spot asteroids that are far away, and they can do it with better detail. Some telescopes are being designed to see asteroids during the day, when the Sun makes it harder to see them. With better telescopes, scientists can track asteroids more accurately and sooner.

2. Space-Based Observatories: A new way to spot asteroids is by using telescopes in space. These space telescopes are not blocked by the Earth's atmosphere, which can make it harder to see asteroids from the ground. A few space missions are already studying asteroids from space. For example, the NEOWISE mission is using a space telescope to look for near-Earth objects, including asteroids. In the future, even more space telescopes will be launched to help scientists keep track of asteroids more efficiently.

3. Radar Systems: Radar systems are another way to track asteroids. These systems send out radio waves to

bounce off asteroids, and by measuring how long the radio waves take to return, scientists can figure out the size, speed, and location of the asteroid. This technology is already being used, but in the future, radar systems will be even more powerful, allowing scientists to track asteroids with much greater detail.

4. Artificial Intelligence (AI): Artificial intelligence is a type of computer program that can help scientists make better decisions by analyzing large amounts of data. With AI, computers can quickly look at thousands of images of space and spot asteroids that might be too small or too far away for humans to see easily. AI can also

help scientists predict the path of an asteroid and determine if it could hit Earth. This technology is still new, but it's making asteroid tracking faster and more accurate.

5. New Space Missions: Several space missions are planned to study asteroids more closely. One of these missions is NASA's *OSIRIS-REx* mission, which is going to an asteroid called Bennu to collect samples. These samples will be brought back to Earth so scientists can study the asteroid's materials. There is also a mission called *DART* (Double Asteroid Redirection Test), which is designed to test a method for changing an asteroid's path. These missions will

help scientists understand more about asteroids and how we can protect Earth from them.

How Will Scientists Learn More About Asteroids in the Future?

The future of asteroid research is exciting because new tools and missions will help scientists learn more about these space rocks. Here are some ways scientists will learn more about asteroids in the years to come:

1. More Close-up Missions: In the future, space agencies will send more spacecraft to visit asteroids. These spacecraft will go up close to the asteroids, take pictures, and even land

on them. By studying asteroids closely, scientists will learn about their size, shape, and surface features. They can also learn about what materials the asteroid is made of, such as metals, water, and carbon. This is important because the composition of an asteroid can help scientists decide the best way to protect Earth if an asteroid is on a collision course.

2. Studying the Asteroid's Orbit: Scientists will keep track of the orbits of asteroids over time. By watching how they move in space, scientists can figure out if any asteroids are headed toward Earth. They will use this information to predict whether the asteroid's path will change.

Sometimes, an asteroid might pass close to Earth but won't hit it. Over time, scientists will learn how to predict an asteroid's future movements more accurately.

3. Learning About Asteroid Impacts: Scientists are also studying the past to understand the effects of asteroid impacts. By looking at craters on Earth and the Moon, they can learn how big asteroids affected our planet. This will help scientists understand what could happen if an asteroid hits Earth again. They can use this knowledge to prepare for potential impacts and find ways to reduce the damage caused by them.

4. Understanding the Role of Asteroids in the Solar System: Asteroids are like time capsules from the early solar system. They are leftovers from when the planets were formed billions of years ago. By studying asteroids, scientists can learn about the early history of our solar system. They can also learn about the building blocks that created the Earth and other planets. This knowledge can help us understand how life on Earth began and what other planets in our solar system might be like.

5. Finding More Asteroids: As technology improves, more asteroids will be discovered. Right now, scientists know about many of the

larger asteroids, but there are still many smaller ones out there that haven't been found yet. New telescopes and radar systems will help scientists spot these smaller asteroids and track their paths. The more asteroids we discover, the better prepared we will be to protect Earth if one of them poses a risk.

What Is the Long-Term Goal of Asteroid Impact Prevention?

The long-term goal of asteroid impact prevention is to protect Earth from the dangers that asteroids could bring. Scientists and space agencies around the world are working together to make sure that if an asteroid is on a path to collide with

Earth, we can do something about it. Here's how scientists are working toward this goal:

1. Deflecting an Asteroid: One of the biggest challenges in asteroid impact prevention is figuring out how to change an asteroid's path. If a dangerous asteroid is headed toward Earth, scientists may need to redirect it. One way to do this is by sending a spacecraft to push the asteroid away from Earth. This idea is being tested through missions like NASA's *DART* mission. If successful, this technique could save Earth from a potentially deadly impact.

2. Breaking Up an Asteroid: Another option for preventing an asteroid

impact is to break up the asteroid into smaller pieces. While this might seem like a good idea, it is not always the best solution. If the asteroid is too large, breaking it into pieces could cause multiple impacts rather than one big one. Scientists are still studying whether this method would work safely and effectively.

3. Developing Early Warning Systems: A key part of asteroid impact prevention is having an early warning system. By spotting dangerous asteroids early, scientists can give people more time to prepare. Early warnings could also help governments make decisions about evacuations or building shelters. With more advanced tracking systems,

we will be able to see asteroids coming from further away, giving us more time to act.

4. Working Together: Finally, the long-term goal of asteroid impact prevention relies on countries and space agencies working together. The threat of an asteroid impact is something that affects everyone on Earth. By sharing information, resources, and ideas, space agencies around the world can come up with the best strategies to protect the planet. Together, we can prevent an asteroid from causing harm.

Conclusion

The future of asteroid impact studies is full of exciting possibilities. With new technologies and space missions, scientists are learning more about asteroids and how to protect Earth. The long-term goal is to develop ways to track, deflect, or break up dangerous asteroids. By working together, we can ensure that Earth remains safe from the dangers of asteroid impacts for many years to come.

Conclusion: Why We Must Keep Learning About Asteroids

Asteroids are fascinating objects in space, but they also pose a potential threat to Earth. While asteroid impacts are rare, they have happened in the past, and they could happen again in the future. The good news is that scientists are working hard to learn more about these space rocks and develop ways to keep us safe from them. In this conclusion, we will discuss the importance of continued asteroid research, how understanding asteroids can help protect us, and what you can do to stay informed.

The Importance of Continued Asteroid Research

Asteroids are not just rocks flying through space. They are pieces of history that can teach us about the formation of our solar system. By studying asteroids, scientists can learn more about the early days of our planet and how life began. But research on asteroids isn't just about understanding the past—it's also about preparing for the future.

Asteroids could pose a risk to Earth, even though large impacts are very rare. The chances of an asteroid hitting Earth are low, but if one does, the damage could be huge. That's why it's important to keep studying asteroids. The more we know, the better we can protect ourselves. The technology we use to find and track asteroids is getting better, and as it improves, we'll be able to spot dangerous asteroids much earlier. This gives us more time to figure out how to prevent a collision.

In addition to protecting us from potential impacts, asteroid research also has other benefits. For example, understanding asteroids can help us learn more about space resources. Some asteroids contain valuable metals, like gold and platinum, which could one day be mined for use in industries on Earth. Studying asteroids also helps scientists learn about the conditions that could make life possible on other planets. These discoveries could lead to new technologies and advances in science that could improve life here on Earth.

How Understanding Asteroids Can Keep Us Safe

The most important reason we need to study asteroids is to protect ourselves from potential asteroid impacts. Even though an asteroid collision with Earth is rare, it's important to be prepared in case it happens. If we can detect asteroids early, we might be able to stop them from hitting Earth or reduce the damage they cause.

Scientists are working on several ways to prevent an asteroid impact, including redirecting an asteroid's path or breaking it into smaller pieces. However, these methods will only work if we have enough time to act. This is why it's so important to continue improving our ability to track asteroids. With better telescopes, radar systems, and other technologies, we can spot dangerous asteroids much earlier than before. The earlier we find an asteroid that could potentially impact Earth, the more time we have to develop a plan to protect the planet.

Another way asteroid research can keep us safe is by studying the effects of past impacts. Scientists study craters on Earth and the Moon to learn more about how asteroid impacts have affected our planet in the past. By understanding these impacts, scientists can predict what might happen if an asteroid hits Earth again. This knowledge helps us prepare for a potential impact and find ways to minimize the damage.

While asteroid impacts may seem like something far in the future, the fact is that the more we understand about them, the safer we can make our planet. Asteroids are unpredictable, but with the right knowledge and tools, we can reduce the risks and be ready for anything.

What You Can Do to Stay Informed

While scientists are doing a lot of work to track and study asteroids, there are things you can do to stay informed about the topic. Learning more about asteroids and space science can help you understand the risks and how we are preparing for them. Here are some ways you can stay updated:

1. Follow Space Agencies: Organizations like NASA, the European Space Agency (ESA), and other space agencies around the world are constantly researching asteroids. They often release updates on new discoveries, missions, and plans for

asteroid impact prevention. By following these agencies on social media or visiting their websites, you can stay informed about the latest news in asteroid research.

2. Watch Documentaries and Read Books: Many documentaries and books are available that explain the science of asteroids in simple, easy-to-understand language. These resources can help you learn about the history of asteroid impacts, current research efforts, and what scientists are doing to protect Earth. Watching videos and reading books can be a fun way to keep up with the latest discoveries.

3. Attend Science Events: Science museums, planetariums, and observatories often host events and talks about space, including topics like asteroids. These events are a great way to learn more about what's happening in the world of space exploration. You might even get the chance to meet scientists and ask them questions about their research.

4. Learn About Space Missions: There are many exciting space missions underway that are studying asteroids. For example, NASA's *OSIRIS-REx* mission is studying the asteroid Bennu, and the *DART* mission is testing a technique to change an asteroid's path. Following these

missions will help you understand how scientists are working to protect Earth and learn more about these fascinating space rocks.

5. Ask Questions: Don't be afraid to ask questions about asteroids or space research. Whether you're in school, at a museum, or talking to a friend, asking questions is a great way to learn more. Scientists love sharing their knowledge, and there are plenty of resources out there to help answer any questions you might have.

Conclusion: Why It Matters

Asteroid research is important for several reasons. By studying these space rocks, we can learn about the early solar system, the possibility of life on other planets, and even find valuable resources. But perhaps the most important reason to study asteroids is to protect Earth from the risks they pose. By improving our ability to track asteroids, we can give ourselves more time to prevent a potential impact. The more we know about asteroids, the better prepared we will be.

As technology continues to improve, scientists will learn more about these space objects and develop better ways to protect Earth. In the future, we may even be able to deflect or destroy dangerous asteroids before they hit. But for now, it's up to all of us to stay informed and support the research that will help keep us safe.

Remember, asteroid impacts are rare, but they can be dangerous. By learning more about asteroids and staying informed about the latest research, you can be part of the effort to protect our planet.

www.ingramcontent.com/pod-product-compliance
Lightning Source LLC
Chambersburg PA
CBHW071029240526
45469CB00006BD/2150